FINESTKIND O' DAY
LOBSTERING IN MAINE

by BRUCE McMILLAN

Down East Books / *Camden, Maine*

ISBN 0-89272-185-5
Library of Congress Catalog Card No. 83-72421
Printed in the United States of America

Reprint edition 5 4 3 2 1

Down East Books / *Camden, Maine*

U.S. Library of Congress Cataloging in Publication Data
McMillan, Bruce A
Finestkind o' day: Lobstering in Maine.
SUMMARY: A young boy learns the meaning of the fishermen's expression, "Finestkind," when he is a sternman on a lobsterboat.
1. Lobster fisheries—Maine—Juvenile literature. [1. Lobster fisheries] I. Title.
SH380.2.U6M3 639'.54'109741

For
Terry and Brett,
Allison and Ruth,
Enid and Betty,
Eva,
and the memory of Tammy

It was a calm spring morning on McGee Island. The sun hadn't appeared on the horizon yet, but the eastern sky was getting lighter. Everyone on the island except Brett was still asleep. Brett was already up and putting on his life jacket, getting ready for his big day. He was going lobstering, just like the lobstermen in Port Clyde. On Brett's last birthday, his parents had given him his own lobster license. Today would be his first day as a sternman, a helper on a lobsterboat.

Brett tramped over the clam flats to his dinghy. Tammy, his dog, trailed along. She was used to going everywhere with Brett, but now he said, "I'm sorry, Tam, you can't come this time. I'm going fishing. You'll have more fun here chasing the field mice. I'll be back later . . . and maybe I'll have a surprise for you!"

Brett set out for Port Clyde. It was a long way to row and he had to keep up a steady stroke if he was going to get there on time. He rowed past Stone Island, where the dock was piled high with lobstertraps. They had been there all winter. The fishermen only lobstered off Stone during the summer, when the lobsters crawled into shallower waters near the shore to shed their shells and grow larger ones. During shedding season these waters would be dotted with buoys marking traps, but now it was clear rowing for Brett.

Once past Stone he was on the open water. He rowed and rowed, dipping his oars in the water, pulling hard on them, lifting them out, and pushing them back, then dipping and pulling again. This was the first time he had rowed by himself all the way to Port Clyde. Brett's father had told him that when he could row four times around McGee Island, he could row to town. One day last week he'd rowed five times around McGee.

Finally Brett reached Blubber Butt Point. Once around the Point he would be in the harbor.

Brett glided by the *Laura B.,* the ferryboat to Monhegan Island. Later in the day she'd be loading up with passengers, supplies, and mail for the island. The trawlers were tied up to the docks. Soon they'd be going out trawling, dragging their fishing nets for cod or haddock or cusk. They caught a lot of fish, but Brett always heard the older fishermen say it wasn't as good as the old days.

Brett was tired from the long row, but he was almost there and he pulled hard on the oars. Brett was going lobstering with Allison Wilson. As he docked and tied his dinghy, he could hear the converted automobile engine chugging away in Allison's lobsterboat.

"Mornin', Brett."

"Good morning, Mr. Wilson," Brett replied.

Allison frowned and said, "I'm sorry, but you can't be my sternman today . . ." Brett tried not to look too disappointed. Then a smile spread over Allison's face. ". . . Unless you call me Allison."

"Okay, Allison!" Brett said. He hopped on board the *Ruth M.* and they were off.

As they were heading out of the harbor Brett asked, "Why did you name your boat the *Ruth M.?*"

"Well," Allison replied, "on lobsterboats it's the custom to name 'em after your wife. The colors are also kind of a custom, white with buff decks and red bottom."

"But I've seen lobsterboats painted all sorts of colors," Brett said.

"Yes," Allison explained, "some of the fellows like the bright colors. A lot of things change with time."

Once they were out of the harbor Allison took an empty bait barrel, turned it over, and set it by the wheel for Brett to stand on. At first Allison kept his hand on the wheel, but then he let Brett steer by himself for a while. After heading out for an hour, Allison slowed the *Ruth M.* to a stop next to one of his lobster buoys. He put on his oilskins, and they were ready to pull the first string of traps.

Allison reached out with his gaff, a pole with a huge fishhook on the end. As he caught the buoy line, he said teasingly, "Oh, oh, I've got a big one. I can tell by the pull that we've got four lobsters in this pot. How many do you think, Brett?"

"Two," guessed Brett.

The water spun off the line as it passed over the pulley, through Allison's gloved hands, around the turning hydraulic wheel, and into the boat. These were deep-water traps, so there was a pile of pot warp on deck before Brett could see the trap coming up, alongside the boat.

With a splash the trap broke the surface. Brett could see spiny sea urchins stuck to the outside laths. Inside he could see something flapping. Allison slowed and then stopped the pot hauler, grabbed the lobstertrap, and pulled it on board.

"How many lobsters?" asked Brett as Allison unlatched the trap. Allison pulled out the first two, which were holding on to each other by their claws, and said, "We've got four in here. I guessed four and you guessed two, but it looks like we both might've guessed right. Let's measure them and see."

Brett was puzzled, but he didn't say anything as he watched Allison get out his lobster measure.

Allison grabbed one of the smaller lobsters. "If you hold it by its back it can't reach around and grab you with a claw. Now you watch how I measure it," said Allison, "and then you measure the other three." He put the end of his measure in the lobster's eye socket. The measure didn't touch the eye, so it didn't hurt the lobster at all. Then he slipped the middle of the measure over the edge of the lobster's back. The body shell just fit inside the measure. It wasn't quite 3-3/16 inches long. Even though it was close, it was a short. So over the side and back into the ocean it went.

"We'll let it grow," said Allison. "After it sheds its shell this summer and grows a new, bigger shell we'll catch that lobster again. Then we'll keep it."

Brett picked up one of the other lobsters, the way Allison had showed him, and measured it. The lobster's body shell was a bit too long to fit inside the measure. It was a keeper. Brett measured the other two. One was a keeper and one a short.

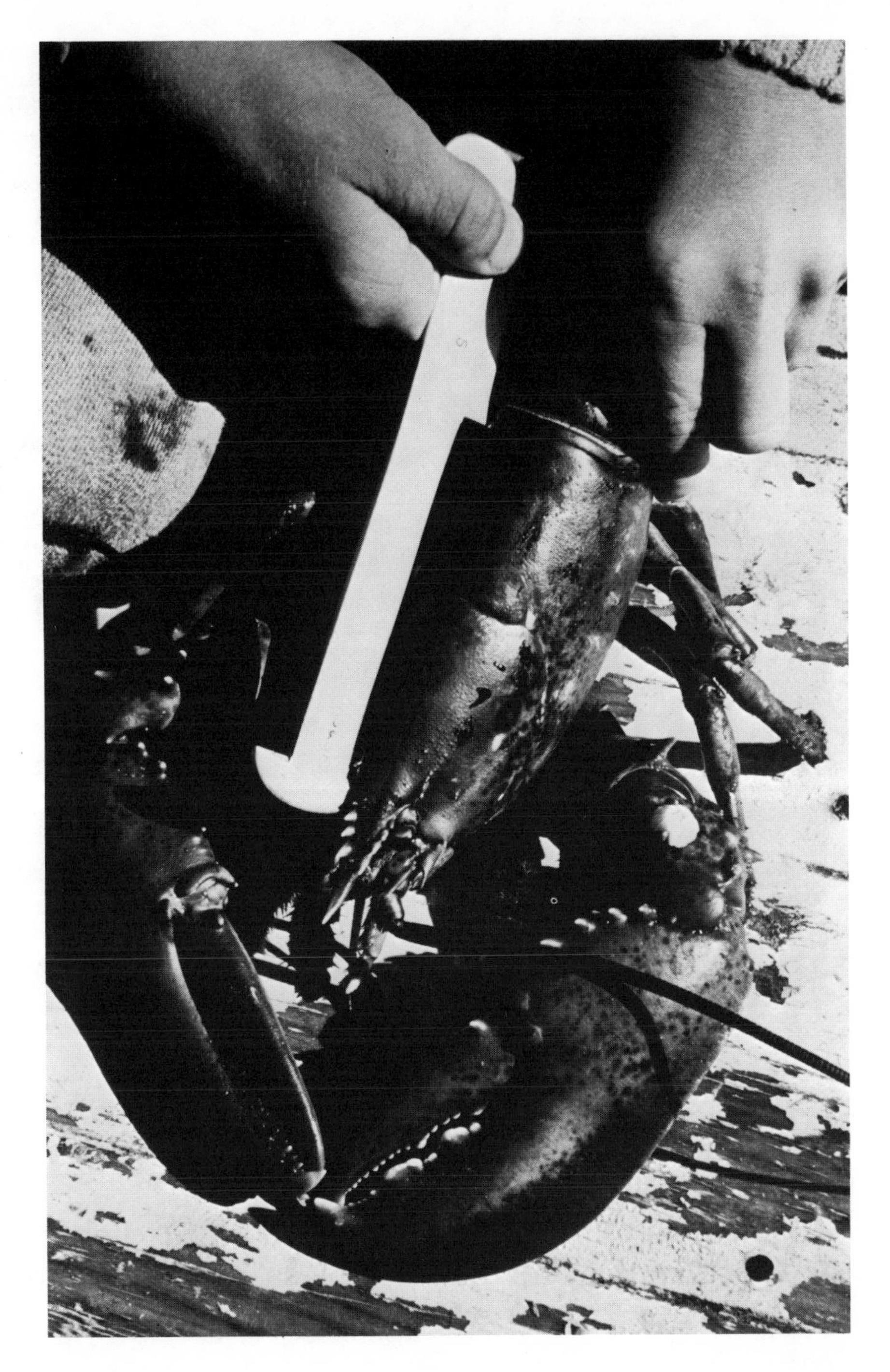

Allison tossed the second short overboard and said, “See, Brett, we both guessed right. We had four lobsters in that trap but only two were keepers.”

Using an expression he'd heard all the fishermen use, Brett replied, “Finestkind!” The fishermen said this when everything was just right, when things were the finest kind.

Allison tossed what was left of the bait in the traps overboard. Brett hadn't seen any sea gulls all morning, but when that old bait hit the water gulls seemed to come from nowhere. They dived into the wake of the boat after the scraps while Allison throttled the *Ruth M.* to the next buoy. The gulls followed along, waiting for the scraps from the next trap.

All morning long Allison and Brett hauled traps. Each time they hauled one they'd try to guess how many lobsters would be coming up. Allison would measure the lobsters and toss the shorts overboard. He passed the keepers to Brett. It was Brett's job to plug the lobsters' claws.

Brett plugged each crusher claw with a peg, then put the lobster into a basket. The ripper claw, the one with the tiny "teeth," didn't need to be plugged because it wasn't as strong as the crusher claw. "With the crusher claw pegged," Allison explained, "they won't be able to fight and pull off each other's claws."

The bait in the traps had attracted other animals besides lobsters. There were spiny green sea urchins that stuck to the traps. There were purple starfish. In one trap they found a hermit crab. "How can that be a crab?" asked Brett. "It looks like a seashell."

"You take it and look inside," said Allison. "Once the shell was empty. But a hermit crab crawled in and made it home."

Brett set the shell down. The crab came partway out and crept along, dragging his home with him.

"Take him back to the island with you and put him in your harbor. He'll be your pet. What are you going to name him?" asked Allison.

Brett thought a minute and then said, “Herman, Herman the Hermit.”

After a while Herman crawled back inside his shell, where no one could see him.

The next pot they hauled had a lobster in it that didn't look like the others. "Why does he look different?" Brett asked.

"You mean she," said Allison. "This one's a female, and she's 'berried.' That means she's loaded with eggs.

"Look close, Brett, and you can see how the eggs stick to each other and to the swimmerets on the bottom side of the mother's tail section. When the baby lobsters hatch they'll look more like tiny shrimp than lobsters. They'll be so small and light they'll float to the surface of the water. After they grow for about ten to thirty days and shed their shells four times, they'll look more like

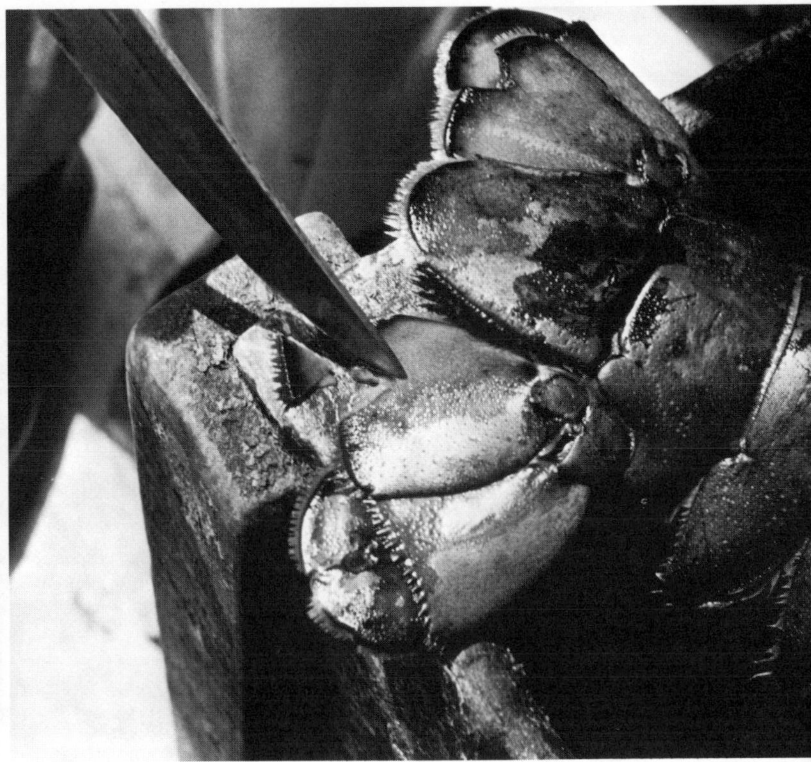

lobsters. Then they're big enough to start living on the ocean bottom, but it will take about seven years before they're big enough to become keepers. Most of 'em won't make it though. When they're tiny and floating around they eat each other or get eaten by fish or the gulls. Some get killed by pollution. All sorts of things can happen to 'em. It takes lots of eggs for there to be enough grown lobsters to make fishing good. That's why we don't keep the females, even if they're big enough. Back over she goes, after I notch her."

With his knife Allison cut triangular pieces out of two of the lobster's tail flaps. "Why did you do that?" asked Brett.

"In case she gets caught again but doesn't have any eggs showing under her tail," Allison explained. "A lobsterman who sees the notches will know she's a female and toss her back."

By late morning they had pulled the last of the winter traps and piled them high on the stern. Allison would repair them in his shop and then set them with shorter lines closer to shore for the summer season. In the summer he'd be hauling his traps once every two or three days instead of once a week.

"Why did you bring bait if we aren't going to set the traps?" asked Brett.

"Well," said Allison, " 'cause we may be through lobstering for the day, but we're not through fishing!"

Allison pulled next to a buoy that was painted his colors but was different from the lobster buoys. One end of his halibut trawl was tied to it. “This trawl has a hundred fishhooks, one every twelve feet, and it’s been settin’ here two days. We may just catch ourselves a hundred-pound halibut on one of these hooks.”

“Oh boy!” exclaimed Brett.

Allison started hauling the trawl line. He baited each hook that was empty and laid out the baited line so as not to tangle it. If they could get a good-sized halibut it would be a good catch. There were only a dozen hooks left to haul when Allison felt a fighting tug and grabbed the line tighter.

Brett was ready with the gaff to help pull the fish aboard. They didn't need it though. The fish was only a twenty-five-pounder and Allison was able to flip it into the boat with a good jerk.

"Oh, Brett, I guess the big one got away," Allison said, and he told Brett about the one he'd almost caught last time. "It was this long," he said, stretching out his arms as far as he could.

Brett wasn't sure if Allison was kidding him and he frowned doubtfully.

"Brett, you can believe me. That's a real fish story," said Allison, winking, as he stood up. Then the fish story teller played out the baited halibut line.

With a small halibut and a good catch of lobsters on board, Allison headed the *Ruth M.* back to port. "Here, Brett, you steer while I clean the halibut and wash down the decks." Standing on the empty bait barrel, Brett took the wheel. "Keep on this course 'til you come to the lighthouse on Marshall Point, then we'll change course."

Allison kept an eye on Brett's course while he cleaned up the boat. He was busy finishing the last of his chores when he heard Brett call back, "There it is!"

As they entered the harbor Allison throttled the boat to a slower speed. When they passed the General Store, Brett remembered Tammy and smiled, thinking about the surprise he'd get for her.

They docked and Brett thanked Allison for letting him work as his sternman. Brett had started off to his dinghy when Allison called him back.

"Why, Brett, you're in such a hurry you've forgotten Herman the Hermit. Here's Herman and here's a dollar for working as my sternman today."

Then Brett rowed off in his dinghy while Allison took the *Ruth M.* to Blaine's dock to leave his lobsters and to gas up.

Before going home, Brett had some stops to make. He rowed by the town docks to the Port Clyde General Store. With Tammy in mind and his dollar in hand, Brett went in.

"Why hello, Brett," said Eva Cushman, the storekeeper. "I talked with your mother on the two-way radio and she told me you went fishin'. She said Tammy's been lookin' all over the island for you."

"I thought she might miss me," said Brett. "That's why I'm getting her a surprise." Brett bought some penny candy and a box of Tammy's favorite dog biscuits.

"Here you go, Brett, and you still have a quarter left. Now you be real careful rowin' home. We want to get those dog biscuits to Tammy."

Brett climbed the hill to the post office to pick up the mail. "Well, hello there!" said Enid Monaghan, the postmistress.

"Hi." said Brett, "I've been lobstering." He told Mrs. Monaghan and her assistant, Betty Wilson, all about his day.

"Why don't you finish your fishing trip by finding my fish?" Mrs. Monaghan said. She always wore a fish, and Mrs. Wilson wore a butterfly. Sometimes they were hard to find, disguised in the print of a dress or blouse.

"It's an easy one today!" Brett exclaimed. "They're on your necklaces."

"I guess we'll have to make it harder next time," said Mrs. Monaghan. "Now you be careful rowing home, and here, don't forget your dog biscuits."

Brett hiked back to the dock where his dinghy was tied up. He knew Tammy would like her treat after a tiring day of chasing field mice. He'd had no trouble picking her favorites because the box had a picture of Tam on it.

Brett started his long row home. The dock where the *Laura B.* had been tied was empty now. The ferry was out at Monhegan, probably loading for the return trip. Brett rowed out of the harbor.

He passed Blubber Butt Point and rowed the long distance across the open water. The tide had been low at sunrise, but now, six hours later, it was high. Brett was able to pass close to the dock on Stone Island and take the shortcut around McGee Island. At high tide he could row over places where he could walk at low tide.

The closer he got to home, the more excited he got.

Tammy was waiting and watching for Brett. As soon as she saw him, she started howling and barking, and then SPLASH! Tammy was swimming out to Brett, wagging her tail while she paddled through the cold ocean water.

"Oh, Tammy," said Brett, pulling her into the dinghy, "you couldn't wait on shore for your surprise, could you?" She sat and shivered while Brett opened the box. Before long Tammy was full of surprises.

As they rowed to shore, Brett found a good spot for Herman and dropped him overboard. "See, Tam? Now Herman will live in McGee Harbor and crawl around the bottom, dragging his shell home with him."

It was a tired Brett and a dog-biscuit-filled Tammy who reached the island's shore. Brett's mother and father had heard Tammy's howling and were coming down the path. Brett's dad yelled down, "How was the fishing?"

And Brett, feeling just like a Port Clyde lobsterman, yelled back, "FINESTKIND!"